This book is dedicated to Link

Table of Contents

Foreword

Did you find this book because you just got your first 3D printer as a gift? Maybe your nephew showed you the part he printed at school and that inspired you to get your own. Or maybe, just maybe, you're a 10-year industry veteran who still busts through every piece of printing content they can find in hopes of learning just a little more about this awesome, quickly developing technology.

However you got here, this book is for you. It's made for beginners and industry experts alike. It's written with the mindset that it can be referenced whether you're using the $200 bed-slinger you got for Christmas or the $250,000 coreXY 1m^3 machine your company got last month.

There's a glossary of common industry terms at the back of this book in case I'm using any that are new to you. If I'm talking about concepts that you've never seen before, keep the faith and read on, as we'll be discussing them in time.

If you're new to this technology, you're making a great decision. You'll still be considered an early adopter. It's only a matter of time before 3DP parts are good enough to be sold in stores or someone can type a description of an object, click go, and have it in their hands an hour later without any tinkering.

The way I most like to describe it is that a hobbyist 3D printer will one day be like an espresso machine. People that like espressos or have a use for one can get it, but it's still affordable and as easy to use as a normal coffee machine. For the moment we can consider 3D printing to be a hobby rather than just a tool, meaning it still requires troubleshooting as well as some trail and error to get to the result you want.

The term 3D printing includes a wide variety of technologies, from fused deposition molding (FDM), stereolithography (SLA), selective laser sintering (SLS), electron beam melting (EBM), binder jetting (BJ), powder bed fusion (PBF), vat-polymerization (VP) and the list goes on. Why so many? Because each has their own strengths and weaknesses in the context of material choice, part geometry, and speed. The starting point for most joining the industry is FDM/FFF, and that's where we'll start as well.

Before we get into the content you came here for, here's my background to give some credibility to what you'll read. If you could care less, here's your queue to jump into the next chapter:

My background focuses on general manufacturing, with a specialty in additive manufacturing and the majority of my experience fitting in FDM/FFF. My love for FDM started 8 years ago, when through my university we ran a project on design tolerancing to create a 3D

printed fidget spinner that we could press fit a bearing and weights into. The project was quick, and the second I had the part in my hands I was enamored. I ordered my own Creality machine to run in my dorm that day. By the time I'd graduated I had started my design/prototyping/consulting firm Burn Social, had 8 FDM machines and two DLP printers, and two employees to help market and run the machines. Since then I've spent a few years working at a $1m^3$ FDM company on the forefront of the industry, and at the time of writing this I run a manufacturing department in the entertainment industry that uses a combination of CNC, 3DP, metrology, etc to make movie magic.

I love to teach almost as much as I love to learn, so I'm excited to share all I've learned in my career with you, the reader. That being said, let's get started!

The History of FDM 3D Printing

Intro

Knowing how something came to be is helpful in predicting where it will go. It brings an appreciation for all it's taken to get to this point. This point being where broke college kids can create useless plastic doo-dads to give out as late birthday gifts, or the basement inventor can prototype his next invention in a matter of hours without ever leaving their workshop. The future is now, so let's see how we got to this point before we dive into the industry today.

The Beginning of a New Era

Imagine a world where you can design and create objects right in your home or office. That's what Fused Deposition Modeling (FDM) 3D printing has made possible! It all started in the 1980s with an engineer named Scott Crump. He came up with a cool idea: a machine that could build things layer by layer, using a special plastic. This was the start of FDM 3D printing.

The First Steps

In the late 1980s, engineer Scott Crump, later joined by his wife Lisa, invented Fused Deposition Modeling (FDM). They patented this 3D printing method in 1989 and founded Stratasys. By 1990, Stratasys had begun commercializing FDM, marking a significant milestone in additive manufacturing and setting the stage for future innovations in 3D printing

DIY and Open-Source Revolution

Fast forward to the early 2000s, and enter the RepRap project by Adrian Bowyer. This project was all about making 3D printers that could replicate themselves! It was open for everyone to contribute, sparking a DIY (Do-It-Yourself) revolution in 3D printing.

Material Matters

As time went on, people started experimenting with different materials for FDM printing. They tried plastics mixed with metal, wood, and even carbon fibers. This opened up a whole new world of things you could make with a 3D printer. We'll talk more about materials later.

A Boom in Popularity

When some key patents expired in the early 2010s, more companies could make FDM printers, which became cheaper and easier to use. Printers like the MakerBot and Prusa i3 became popular, and more people started using them, not just in industries but also at home.

FDM in Action

FDM printers are now used for making quick prototypes, tools, and even final products. They've been a game-changer in the medical field, helping make things like prosthetic limbs and surgical guides.

Overcoming Challenges

Like any technology, FDM has its challenges. People have been working on making it faster, more accurate, and giving it a better finish. Innovations continue, making FDM printers smarter and more capable.

Eco-Friendly Future

FDM is also stepping up in sustainability. There are now filaments (the plastic material used in FDM) that are biodegradable and recycled. This is helping reduce waste and could lead to more local, eco-friendly manufacturing.

The Ongoing Journey

FDM 3D printing started as a simple idea and has grown into a technology that's changing how we make things. It's a story of creativity, collaboration, and innovation. With continuous advancements and a growing community of enthusiasts, FDM's future looks bright and full of possibilities. If you just got your first machine, welcome! If you're an industry veteran, thank you for helping grow our world and kudos to taking the steps to continue your journey!

With this brief history laid out, let's get into what machines look like today in this next chapter.

Types of FDM 3D Printers

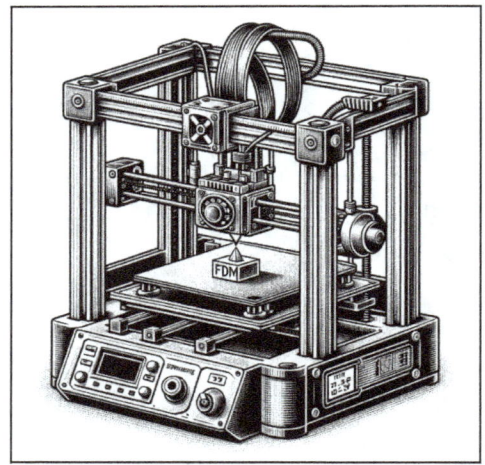

Bed Slingers

Description: The most common hobbyist machine to date, bed slingers operate with separate X, Y, and Z axes each with their own stepper motors. The X typically carries the printhead and moves up and down on the Z, while the Y 'slings' forward and backwards as one piece.

Pros:
- Widely available
- Cheap to make
- Easy to operate
- Good for beginners

Cons:
- Limited printing speed,
- More moving parts,
- May have difficulty with overhangs and bridges

Material Range: PLA, ABS, PETG, TPU, Nylon, and various specialty filaments. "Hobbyist grade" materials through low-grade engineering materials

Example Parts: Prototypes, figurines, household objects, aesthetic parts

Companies: Creality, Anycubic, Prusa Research.

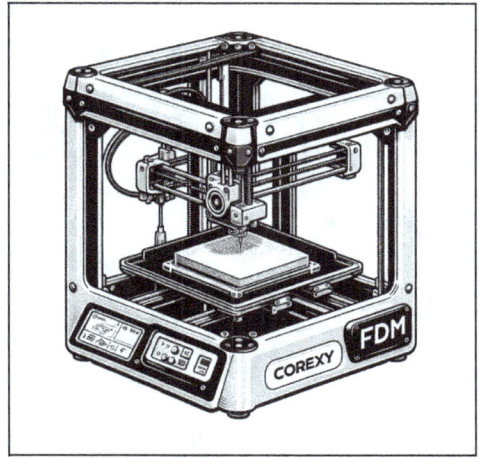

CoreXY

Description: CoreXY printers use a belt-driven system to move the print head in the X and Y directions, while the bed moves down the Z on lead screws

Pros:
- Fast and precise
- Reduced vibrations
- Easy to enclose
- Higher print quality
- Larger build volumes

Cons:
- More complex design,
- May be harder to calibrate and troubleshoot.

Material Range: Hobbyist grade materials (not great with TPU) Capable of printing PEEK/PEI with motors housed outside enclosure

Example Parts: Detailed architectural models, functional prototypes, cosplay props.

Companies: Bambu Labs, Makerbot, RailCore Labs.

Burn
Social

Delta FDM 3D Printers:

Description: Delta 3D printers use three vertical columns and arms to move the print head in a circular motion. This design allows for rapid movements and is often used for taller prints.

Pros:
- Fast printing speeds
- Tall build volume
- Easy to enclose
- mesmerizing to watch

Cons:
- Less intuitive for beginners
- harder to calibrate
- limited for certain geometries

Material Range: Most hobbyist filaments outside TPU. Low-grade engineering filaments.

Example Parts: Tall sculptures, vases, architectural models, lampshades.

Companies: Anycubic, FLSUN, SeeMeCNC.

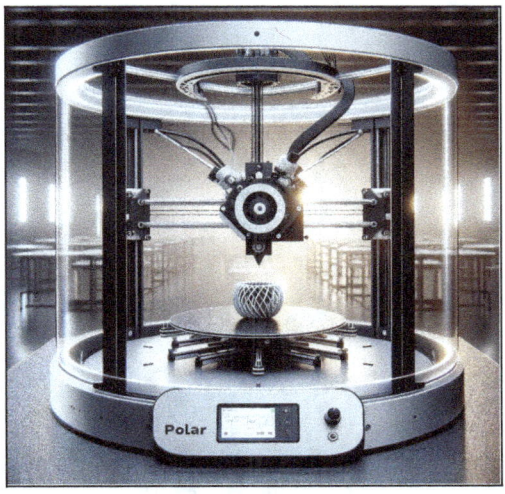

Polar FDM 3D Printers

Description: Polar 3D printers are a less common design that uses a rotating print bed and print head, often resembling a turntable. The print head moves in a circular motion around the fixed center point.

Pros:
- Smooth and continuous motion,
- suitable for round-shaped objects simple design

Cons:
- Slower printing speed
- limited adoption compared to other designs

Material Range: PLA, ABS, PETG, and some specialty filaments.

Example Parts: Round objects like containers, artistic sculptures.

Companies: Polar 3D (no longer in production), RoVa3D (Rapid Prototyping 3D Printer).

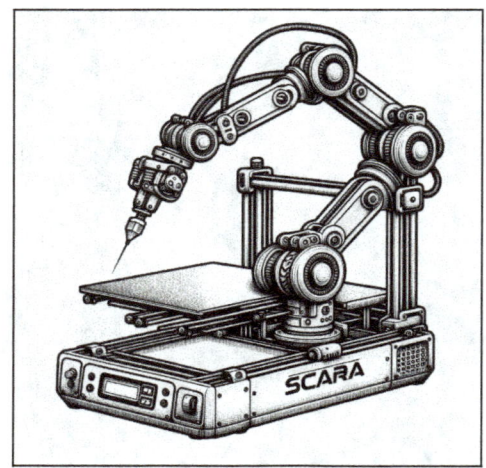

SCARA FDM 3D Printers:

Description: SCARA (Selective Compliance Articulated Robot Arm) 3D printers use an articulated arm to move the print head. This design is more commonly found in industrial settings but is used in some 3D printers.

Pros:
- Fast and precise movements
- Suitable for complex geometries

Cons:
- Less common in consumer printers
- Higher cost
- Harder to program
- Hard to enclose

Material Range: Hobbyist and limited engineering-grade filaments.

Example Parts: Custom molds, industrial prototypes, complex mechanical parts.

Companies: Caracol, nScrypt (manufactures hybrid SCARA systems).

Note

The examples provided are based on typical use cases, and the material range may vary depending on the printer's specific capabilities, modifications, and ambient workspace. Always refer to the manufacturer's specifications for accurate information on material compatibility and part examples. Additionally, the 3D printing industry is continuously evolving, and new printer types and designs will emerge over time.

Filament Types

PLA (Polylactic Acid):

Description: PLA is one of the most popular and widely used FDM 3D printing materials. It is derived from renewable resources like cornstarch or sugarcane, making it biodegradable and environmentally friendly.

Pros:
- Easy to print,
- Minimal warping,
- High accuracy
- Low odor
- Good for beginners.

Cons:
- Relatively brittle
- Limited temperature
- resistance.(~90*F)

Applications: Prototypes, artistic models, tight tolerance

Bed Adhesion: BuildTak, PEI Sheet, Glass Bed with adhesive options like PVA glue stick or painters tape

ABS (Acrylonitrile Butadiene Styrene)

Description: ABS is a durable and impact-resistant thermoplastic. It offers better temperature resistance compared to PLA but requires a heated print bed and an enclosed printing environment to minimize warping. It also needs ventilation, as it emits VOCs.

Pros:
- Strong and tough,
- Higher temperature tolerance.

Cons:
- Prone to warping,
- Toxic odor during printing.

Applications: Functional parts, automotive components, LEGO-like models.

Bed Adhesion: BuildTak, Textured or Powder-Coated PEI, Enclosure or Printer Hood with adhesive options like PVA glue stick or ABS glue (ABS dissolved in acetone)

PETG (Polyethylene Terephthalate Glycol-Modified)

Description: PETG is a hybrid material that combines the best features of PLA and ABS. It is known for its toughness, clarity, low warping, and resistance to chemicals. A big feature is being able to print water-tight parts with the right temp settings.

Pros:
- Impact-resistant,
- Chemical resistance,
- Easy to print.

Cons:
- Slightly flexible,
- May not be suitable for high-temperature applications.
- Poor Support Removal

Applications: Containers, functional parts, water bottles.

Bed Adhesion: PEI Sheet, G10, or Glass Bed with adhesive options like PVA glue stick.

Burn Social

TPU (Thermoplastic Polyurethane)

Description: TPU is a flexible and elastic material commonly known as a rubber-like filament. It is known for its excellent flexibility and shock-absorbing properties. TPU can be found in a range of shore-hardness depending on the desired application.

.

Pros:
- Flexible and elastic
- Resistant to abrasion.
- Chemical resistance

Cons:
- Challenging to print (Direct drive setup is best)

Applications: Gaskets, phone cases, vibration dampeners.

Bed Adhesion: BuildTak, G10, or Textured PEI with adhesive options like glue stick.

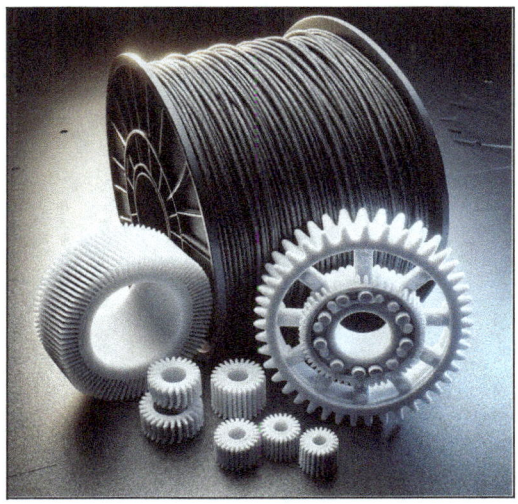

Nylon (Polyamide)

Description: Nylon is a strong and durable engineering material with excellent mechanical properties. It absorbs moisture, so it requires proper storage and drying before printing. It's best to use an enclosure when printing Nylon

Pros:
- High strength, toughness, and durability

Cons:
- High moisture absorption
- Prone to warping

Applications: Gears, mechanical parts, functional prototypes.

Bed Adhesion: PEI Sheet or Glass Bed with adhesive options like glue stick and adhesion promoters like isopropyl alcohol.

Burn Social

PC (Polycarbonate)

Description: PC is a high-temperature thermoplastic known for its exceptional strength, impact resistance, and optical clarity. It requires a heated chamber to reduce warping and layer separation during printing.

Pros:
- Excellent mechanical properties
- High-temperature resistance.

Cons:
- Requires a heated chamber
- Can be challenging to print.

Applications: Engineering parts, automotive components, safety equipment.

Bed Adhesion: Heated Chamber, Enclosure or Printer Hood, PEI Sheet with adhesive options like glue stick.

HIPS (High Impact Polystyrene)

Description: HIPS is similar to ABS in terms of properties but is mainly used as a support material due to its solubility in limonene and ease of snapping.

Pros:
- Good for support material
- Easy to print.
- Leaves clean contact surfaces

Cons:
- Can only print with materials that have similar thermal properties
- Emits an odor during printing

Applications: Best used on materials that don't separate easily (ie PLA, PETG, PC, etc)

Bed Adhesion: BuildTak, Textured PEI, G10, or Glass Bed with adhesive options like glue stick.

**Burn
Social**

PVA (Polyvinyl Alcohol)

Description: PVA is a water-soluble support material compatible with dual-extrusion 3D printers. It dissolves in water, leaving clean prints without manual support removal.

Pros:
- Water-soluble for easy support removal
- Leaves cleanest contact surfaces

Cons:
- Can be sensitive to humidity
- May require additional drying

Applications: Complex geometries, overhangs, and internal structures that can't be reached

Bed Adhesion: BuildTak, Textured PEI, G10, or Glass Bed with adhesive options like glue stick and adhesion promoters like isopropyl alcohol.

Wood-filled Filaments

Description: Wood-filled filaments are composite materials with PLA or other base polymers infused with wood shavings. They offer a natural, wood-like appearance to printed objects and can be colored with wood stain.

Pros:
- Aesthetically appealing
- Sandable/tappable

Cons:
- Limited mechanical properties
- May require special nozzles
- Clogs easily

Applications: Artistic and decorative prints, wooden models.

Bed Adhesion: BuildTak, Textured PEI, G10, or Glass Bed with adhesive options like glue stick.

Burn Social

Metal-filled Filaments

Description: Metal-filled filaments are a base polymers infused with metal particles (e.g., copper, bronze, iron). They provide a metallic appearance and enhanced weight to printed objects.

Pros:
- Metallic finish without fullmetal printing
- Easy to handle.

Cons:
- Limited metal properties
- May require special nozzles

Applications: Jewelry, sculptures, decorative objects.

Bed Adhesion: BuildTak, Textured PEI, G10, or Glass Bed with adhesive options like glue stick.

Carbon Fiber-filled Filaments

Description: Carbon fiber-filled filaments are PLA or other base polymers infused with carbon fiber particles. They offer increased strength, stiffness, and weight reduction in printed parts.

Pros:
- High strength-to-weight ratio
- Enhanced mechanical properties

Cons:
- Abrasive to nozzles
- Require upgraded nozzles and extruder motor gear

Applications: Lightweight structural parts, functional prototypes.

Bed Adhesion: BuildTak, Textured PEI, G10, or Glass Bed with adhesive options like glue stick.

Burn Social

Conductive Filaments

Description: Conductive filaments are infused with conductive materials like graphene or carbon, providing electrical conductivity to printed parts.

Pros:
- Electrical conductivity for functional applications.

Cons:
- Limited conductivity compared to metal, specialized applications.

Applications: Circuitry, sensors, EMI/RFI shielding.

Bed Adhesion: BuildTak, Textured PEI, G10, or Glass Bed with adhesive options like glue stick.

Burn Social

Glow-in-the-Dark Filaments:

Description: Glow-in-the-dark filaments contain phosphorescent pigments that absorb light and emit a glow in the dark.

Pros:
- Fun and creative
- Easy to print

Cons:
- Limited glow duration
- Less suitable for weight-bearing prints

Applications: Novelty items, toys, and decorations.

Bed Adhesion: BuildTak, PEI Sheet, G10, or Glass Bed with adhesive options like glue stick.

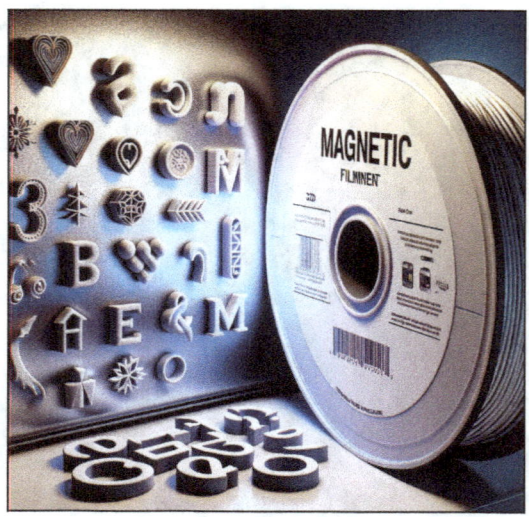

Magnetic Filaments

Description: Magnetic filaments contain iron or other magnetic particles, enabling printed objects to exhibit magnetic properties.

Pros:
- Magnetic properties for interactive prints.

Cons:
- Limited magnetic strength
- Specialized applications.

Applications: Custom magnets, educational models.

Bed Adhesion: PEI Sheet with Magnetic Base, G10, or Glass Bed with adhesive options like glue stick.

Heat-sensitive (Thermochromic) Filaments

Description: Thermochromic filaments change color with temperature fluctuations, revealing a color shift based on ambient temperature.

Pros:
- Fun and artistic prints
- Unique visual effects.

Cons:
- Limited color options
- Specialized applications.

Applications: Artistic prints, visual temperature indicators.

Bed Adhesion: BuildTak, PEI Sheet, G10, or Glass Bed with adhesive options like glue stick.

Burn
Social

Note

It's essential to note that the selection of bed adhesion surfaces, adhesives, and accessories may depend on the specific 3D printer model and the filament being used. Experimentation and adjustments may be required to achieve the best results for each material and print. Always refer to the manufacturer's guidelines and recommendations for bed adhesion and accessory use to ensure successful 3D printing.

As this is a hobbyist/beginner's guide, we skipped over more advanced engineering materials such as PEEK / Ultem (PEI). They're extremely cool and are worth looking into if advanced materials *peek* your interest. They require build chambers that can reach over 100 degrees C at a minimum, which require special machines with some titanium components, special heat shielding, and motors mounted outside the build volume.

Troubleshooting

Troubleshooting - Preventative Maintenance

What's the best way to fix an issue with your machine?

Never having one in the first place!

Below are a few steps you can take to keep your printer running and avoid spending time/money on repairs. Better to use your time designing parts and coming up with new ideas to print!

Input Voltage

Are you reading this from the US, the EU, maybe EA? Depending on your location, your home or office's normal outlets may have a different voltage. For the US, this is typically 120V, while Europe is usually 240V and Japan is 100V. Most machines have a 110V-230V switch depending on which you have, and having this set wrong can be disastrous if not dangerous.

A quick google search will help, but the best way to know is to test with a voltmeter before plugging your machine in.

Surge Protection

Plugging your printer into a surge protector is a great way to not fry your motherboard or sensors. Most printers have fuses in their power

38

switches, but they have absolutely failed and you most likely don't have an extra one on-hand. Better safe than sorry.

Lubrication

Greasing your Z-axis lead screws on a cartesian machine is something we all forget about if we don't have a reminder. Your Z-axis bears the weight of your X-rail and printhead, and you depend on it to make sure your layers are precisely spaced. Dust and debris accumulate on the threading and cause excessive strain on the Z-motor. If it's bad enough, you can also get uneven/skipped layers.

All you need to do is check the spindle(s) for buildup and once the grease is dirty wipe clean with a paper towel/cloth and reapply. The grease varies by the material of the lead screw, so check with your manufacturer but in most cases it's a mineral-based lubricant.

Any ball bearings on your machine are a little easier. Usually a squirt of WD40 can eat away gunk and keep them running smoothly.

Belt Tensioning

Your printer manufacturer will likely have recommendations for a desired belt tension you can follow. There are a variety of phone apps that are accurate enough for a hobbyist machine, and once you've got the belts adjusted a few times you'll get the intuition to do this by hand.

Burn Social

This doesn't have to be perfect. If your belts are too tight, you'll put excess strain on the motor, shortening its life and possibly causing it to skip. Too loose and you;ll see your timing belt skip off its anchor.

Extruder Assembly

An extruder assembly usually consists of sharp teeth on a spring arm to grip filament. With time, the spring will lose its compressive strength and the teeth will wear (especially with abrasive filaments i,e, carbon fiber nylon or aluminum-filled PLA). The spring should last you a while, and if you mainly use prototyping materials the gear will last for years.

The best way to know your assembly is working right is to do a flow-rate calibration (look this up if you're unsure how to on your machine) and to pinch the filament and feel it moving as the extruder gear spins.

The main thing I do regularly is after a few dozen hours of printing or a nasty clog I'll clean the filament from between the extruder gear teeth with a toothpick, helping them grip much better.

Filament Storage

Keeping material dry is crucial for a successful print. Some materials are very sensitive to humidity, such as nylon and TPU, and

therefore should be stored in a closed container with desiccant. This becomes more of a serious issue if you live in a humid climate.

You'll be able to tell your filament is saturated when you hear a crackling and popping when printing (just like wet logs on a fire). This can result in little bubbles in the print, and the moisture in the filament will cause it to expand and possibly clog.

If this happens to you there's no need to worry. Most filament can be dried in a conventional oven, and I personally use a 1kg filament-sized food dehydrator. Definitely do some research on the specific filament brand you're using before doing this, as it can be a mess or even dangerous if temperatures / times are set wrong.

Troubleshooting - Print Issues

Believe it or not, you are able to tell what type of mechanical issues your machine is having just by looking at the parts it's spitting out! This is a diagnostic strategy I took on early in my FDM printing journey that has worked wonders, and I'm sharing it with you here. As long as your machine can move and extrude, these steps work. If you haven't gotten that far yet, start with consulting your machine manufacturer's startup guide.

What you'll find is that troubleshooting can be a bit of a balancing act. You change one setting to fix stringing, and now all of a sudden your part won't stay stuck to the bed! It's a puzzle that you'll find you have to tweak pieces of every so often. Keep asking questions and making changes to your setup/slicer settings and you'll get to where you need to be, Also, do keep this book handy as a reference to help you get back on track quickly. You'll find a notes section at the back of your book to use as a maintenance log if you choose to.

Bed Adhesion Problems

Identification: Your prints refuse to stick to the bed, or even halfway through your print they peel. This can cause your printhead to crash, filament to build up on the heating element, or even if your part magically comes out it won't be correctly shaped

Fix:
- Use BuildTak plate, PEI, magigoo, etc (see bed adhesion recommendations in the "Filament Types" chapter)
- Raise or lower the bed temperature based on material manufacturer requirements
- Lower printing speed, or lower nozzle height slightly
- Raise nozzle temperature to the higher end of recommended manufacturer settings
- Add an enclosure to stop drafts from cooling filament too quickly.

Blobs and Zits (Random Bumps on the Print Surface)

Identification: Small irregular bumps or blobs on the surface of the part.

Fix:
- Optimize retraction settings based on manufacturer recommendations to reduce filament oozing during travel moves.
- Adjust print speed, acceleration, and jerk settings to reduce sudden changes in movement.

Note: One tiny blob typically forms per layer regardless, usually referred to as the "Z Seam" where the part changes layers. Lowering print speed will help make this smaller, but most slicer software automatically aligns this layer change point with sharp corners on the part to hide it.

44

Elephant Foot (Base Compression)

Identification: The base of the print appears flattened or bulging, with the first layer slightly squished.

Fix:
- Reduce the first layer printing temperature and adjust the bed level to prevent over-compression.
- Increase the nozzle distance from the bed for the first layer.
- Adjust the flow rate for the first layer to reduce excessive extrusion.

Note: Careful with lowering the flow too much or printing too far from the bed, it will cause the part not to grip the build plate. See "Bed Adhesion Problems".

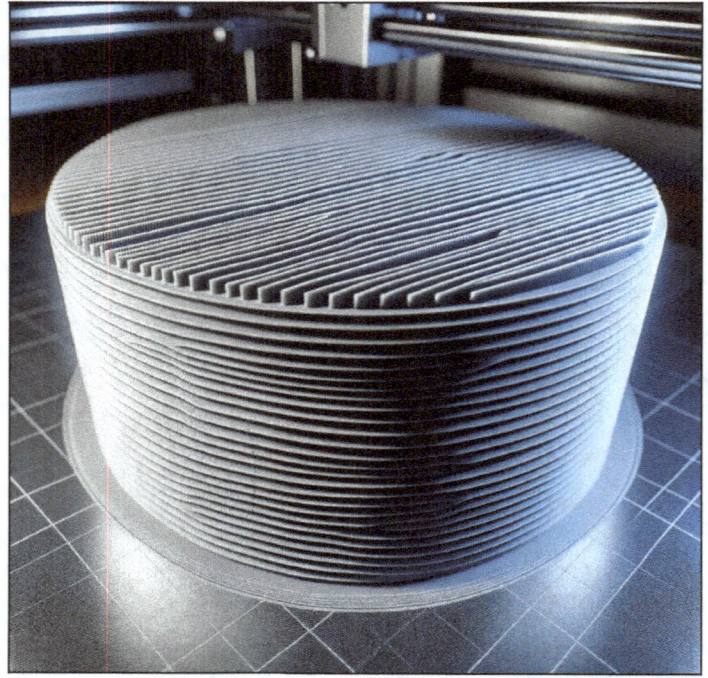

Inconsistent Line Width (Variable Under- or Over-Extrusion)

Identification: The width of printed lines or perimeters varies on different layers, affecting the print's dimensional accuracy.

Fix:
- Perform a flow rate calibrate to dial in the extruder steps/mm. Look for videos online if you are unsure how to.
- Check that extruder motor rotates smoothly
- Ensure consistent filament diameter measurement.

Inconsistent Layer Height

Identification: The print shows variations in layer height, leading to a visibly uneven surface, layer separation, or print in air.

Fix:
- Check for mechanical issues, such as binding or misalignment of the Z-axis lead screws.
- Lubricate Z-axis components to reduce friction and ensure smooth movement.
- Check Z-axis steps/mm (move Z up a specific amount of inches then measure actual movement)

Burn Social

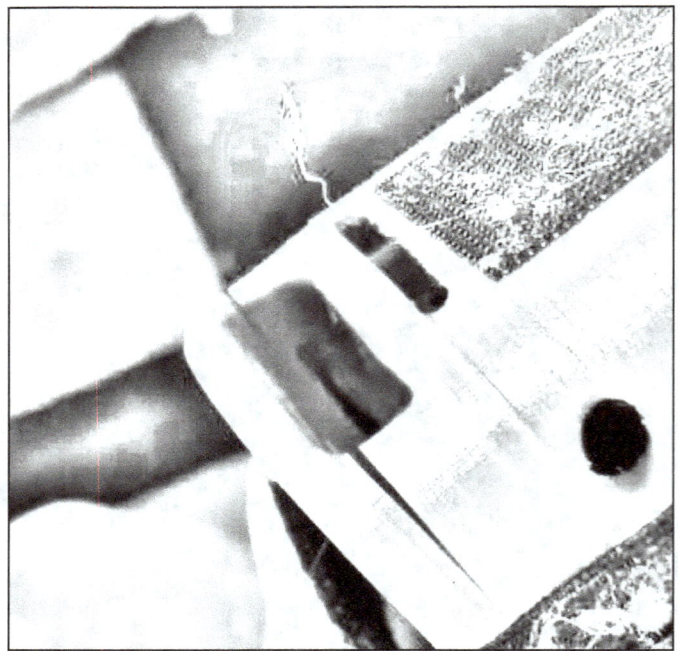

Layer Splitting (Layer Adhesion Problems)

Identification: Layers separate or delaminate, resulting in weak prints that break apart easily.

Fix:
- Increase the printing temperature to improve layer bonding.
- Slow down the print speed to allow for better layer adhesion.
- Ensure the printer's cooling fan is not cooling the print too rapidly, affecting layer adhesion.
- Check the filament for moisture or contamination, and dry it if necessary.

Layer Shifting (Misaligned Layers)

Identification: The layers of the print are not aligned correctly, causing a shift in the print's position.

Fix:
- Check for loose belts on the X and Y axes, and tighten them if needed to counter slippage.
- Make sure the printer's stepper motor drivers are correctly calibrated.
- Confirm that the printer is on a stable and level surface.
- Reduce print speeds or acceleration settings to prevent excessive force on the printer's mechanical components. (especially for direct-drive setups)

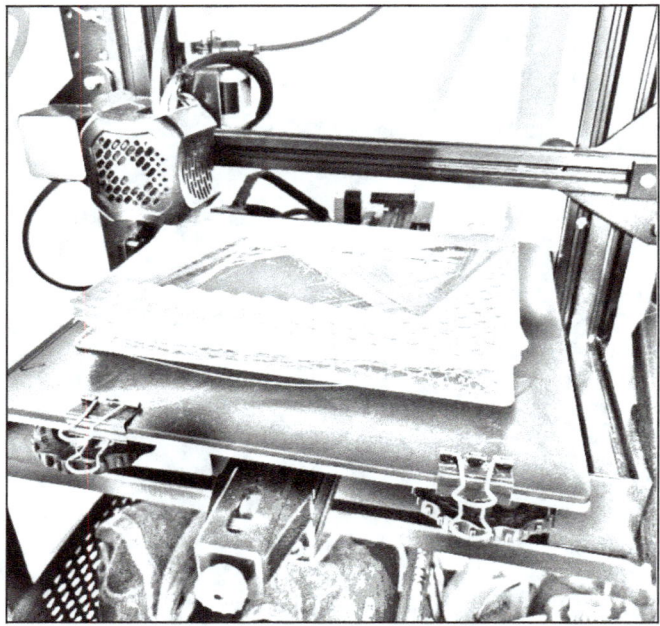

Nozzle Clogs

Identification: The print shows gaps/inconsistent extrusion due to nozzle blockages. Maybe the extruder gear is spinning and nothing comes out

Fix:
- Unclog the nozzle using a nozzle cleaning kit or a cold pull technique (see videos online).
- Print with high-quality and properly stored filament to minimize the risk of clogging.

Note: After you've reseated the nozzle, pull the filament out and check to see if the extruder gear ate into it while it was stuck (it usually does!). Just snap the filament past that point and feed it in again.

Overheating and Layer Drooping

Identification: The print shows signs of overheating, such as melted or sagging layers, especially on overhangs and bridges.

Fix:
- Reduce the printing temperature within manufacturer guidelines.
- Enable a fan to improve cooling for overhangs and bridges.
- Increase the cooling fan speed for problematic areas of the print.
- Print multiple objects simultaneously to allow for more cooling time between layers.

Burn Social

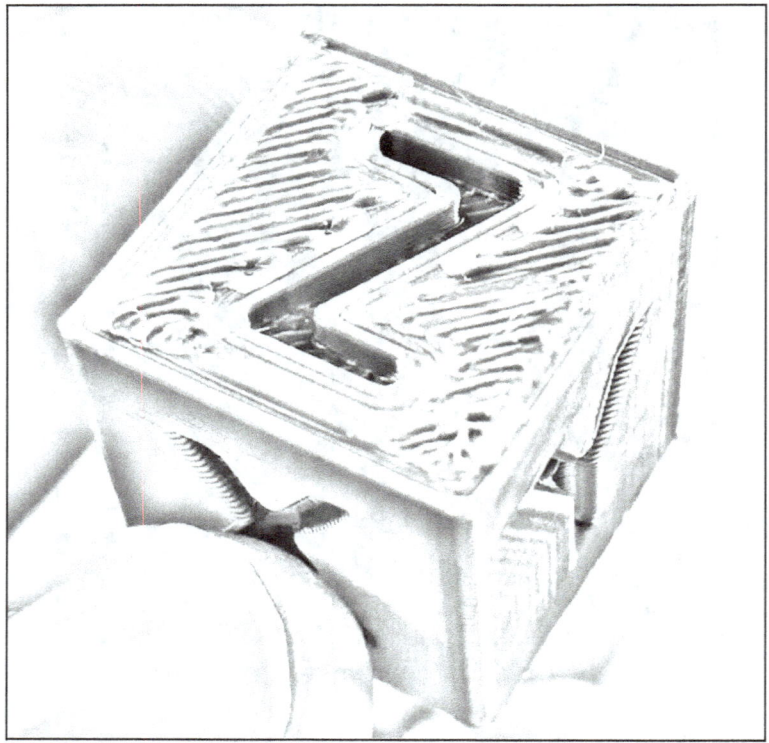

Pillowing (Uneven Top Layers)

Identification: The top layers of the print are uneven or have a puffy appearance.

Fix:
- Increase the top layer thickness in the slicer settings.
- Adjust infill density and pattern to provide better support for the top layers.
- Use more top layers to create a smoother top surface.

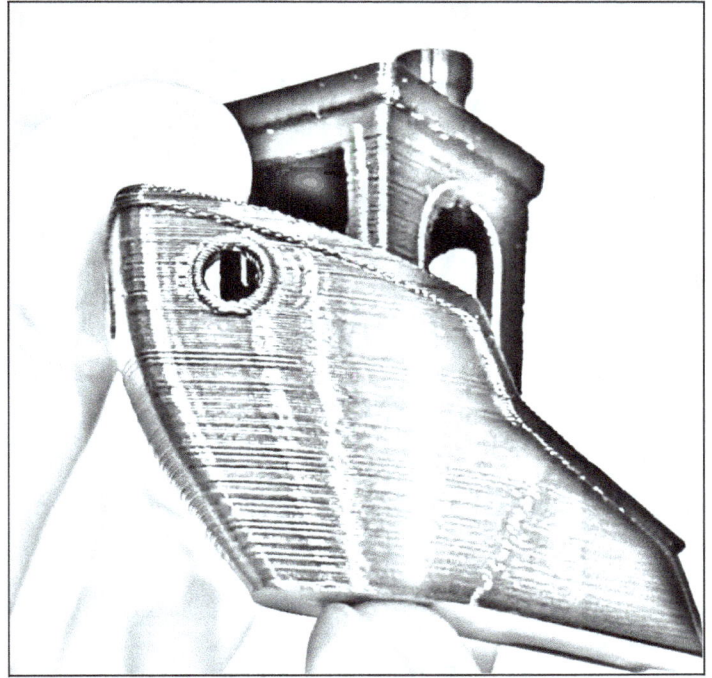

Vibration/Ringing in Part

Identification: Visible lines or patterns giving a waviness to the part surface.

Fix:

- Check for bent or misaligned lead screws or threaded rods and replace them if necessary.
- Ensure that all Z-axis components are securely fastened to minimize vibrations.
- Ensure Y-axis rollers are snug to the Y-rail.
- Confirm the machine is placed on a stable surface
- Reduce print speed if none of the above work.

Skipped Layers

Identification: Sections of the print are missing entire layers, leading to gaps or incomplete structures.

Fix:
- Check for loose belts, pulleys, and/or stepper motor connections.
- Increase the motor current if necessary to prevent skipping steps.
- Grease Z-axis spindle to reduce any friction on it
- Check filament diameter to ensure there are no bulges or thin spots

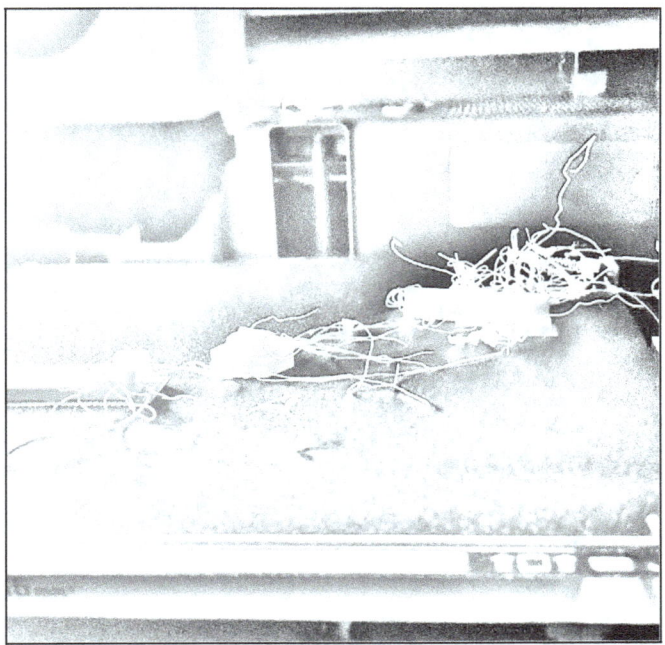

Spaghetti

Identification: Large mess of extruded filament all over bed. Likely a buildup of filament on the heating element as well.

Fix:
- Usually caused by either bed adhesion issues or under-extrusion. See related sections of this chapter.

Note:

After cleaning your bed off, double-check the nozzle is completely clean. If not, clean it at filament printing temperature. Otherwise, there's a high chance that material will drip off onto the next part and cause subsequent issues.

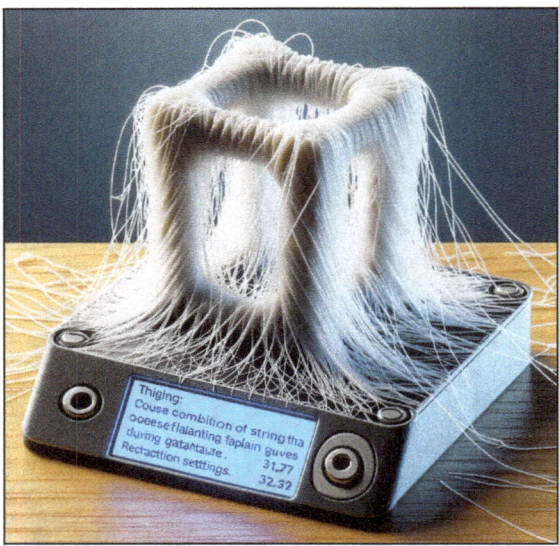

Stringing (Excessive Filament Oozing)

Identification: Thin strands of filament connecting different parts of the print where the nozzle moves between them.

Fix:
- Reduce the printing temperature to minimize filament oozing during travel moves.
- Enable or increase retraction settings in the slicer software to pull back filament when moving between print areas.
- Adjust travel speed and distance to reduce stringing.
- Optimize retraction settings by printing a calibration test.

Note: Strings will shrivel up and be brushed off with a (very) quick pass with a heat gun. I'll usually ignore strings on parts that need to be strong and that aren't aesthetic. It's a compromise for my higher-than-normal temperature and flow settings making the layers bond well.

Burn Social

General Troubleshooting Note

When troubleshooting 3D printing issues, it's essential to consider factors like printer hardware, filament properties, slicing settings, and environmental conditions. Best practice is to change one setting or mechanical component at a time, and keep a log of changes made during troubleshooting to track their impact on print quality.

Additionally, referring to the printer's manual, online forums, and community resources can provide valuable insights and solutions to specific problems. With patience and persistence, your parts can look as good as the ones these manufacturers advertise.

... And Beyond

I hope this little guide has been helpful for taking your first steps into the world of FDM printing, and that it's given you ideas on how you want to move forward.

The 3DP industry is constantly growing and changing. As the technology adapts to help us print bigger/faster/cleaner we'll continue to see new uses for it pop up.

Maybe you'll be the person to bring glass 3D printing to a consumer desktop machine, or make the first combination 3D printer / waterjet cutter. Maybe you're into being the first in line for the newest crowdfunded technology or want to start a youtube channel once you have a novel application to share. Any way you slice it, it's going to be a great time watching all these new aspects of the technology grow.

If you enjoyed this read and have any feedback, feel free to reach out through our website (GoBurnSocial.com) or my email (RHayburn@goburnsocial.com). For video tutorials, check our YouTube channel (@BurnSocial) where I post a lot of short-form calibration and troubleshooting videos, as well as reviews on new machines and materials.

I'm happy to respond to comments, questions, and the like if I can add value to your experience with 3DP. We've barely scratched the surface into what this technology can offer, and if this book gets decent reception I'll be continuing the series with more beginner books and guides that focus on all different types of additive manufacturing.

Happy printing and best wishes on your remarkable journey into the captivating world of FDM 3D printing!

Glossary

ABS: Acrylonitrile Butadiene Styrene - A common thermoplastic filament used in 3D printing.

Adaptive Layer Height: A feature in some slicers that adjusts the layer height based on the level of detail required in different parts of the print.

Bed Adhesion: The ability of the first layer of a print to stick securely to the print bed.

Bed Leveling: The process of adjusting the print bed to ensure it is parallel to the printer's movement.

Bowden Extruder: An extruder design where the motor is separate from the print head.

Bowden Tube: The PTFE tube used to guide filament from the extruder to the print head in Bowden extruder setups.

Brim: A few layers of material printed around the base of the object to improve bed adhesion.

Build Volume: The maximum size of the object that can be printed in a particular 3D printer.

Calibration Cube: A small test print used to calibrate the printer's settings and ensure accuracy.

Cooling Fan: A fan used to cool the printed layers for better print quality, especially for PLA prints.

Burn
Social

Creep: The deformation of a printed object under sustained load or high temperature.

Cross-linking: The process in which thermoplastic materials chemically bond during printing to improve print strength.

Cura: A popular open-source slicing software for 3D printing.

Deburring: The process of removing rough or sharp edges from a 3D printed object.

Direct Drive Extruder: An extruder design where the motor is mounted directly on the print head.

Draft Mode: A print setting that prioritizes speed over print quality for rapid prototyping.

Extruder: The component that pushes the filament through the nozzle for printing.

FDM: Fused Deposition Modeling - A 3D printing process that builds objects layer by layer by extruding melted filament through a nozzle.

Filament: The material used in FDM 3D printing, typically in the form of a spool, such as PLA, ABS, PETG, etc.

Firmware: The software embedded in the 3D printer's hardware that controls its movements and functions.

Flow Rate: The rate at which filament is extruded during printing, affecting print density and strength.

G-code: The language used to communicate instructions to 3D printers, containing commands for movement and extrusion.

Gantry: The structure that holds and moves the print head in a 3D printer.

Heated Bed: A print bed that can be heated to improve print adhesion and reduce warping.

Homing: The process of moving the print head to its starting position before each print.

Infill: The interior structure of a 3D printed object, which can be adjusted to control the print's strength and material usage.

Infill Density: The percentage of the interior volume that is filled with infill material.

Jerk: The rate at which the printer accelerates or decelerates, affecting print quality and ringing artifacts.

Layer Adhesion: The strength of adhesion between consecutive layers in a 3D print.

Layer Height: The thickness of each printed layer, determined by the Z-axis movement.

Linear Advance: An advanced calibration technique used to optimize pressure control during printing, reducing ooze and improving print quality.

Linear Rails: Precision linear guides used in some 3D printers to provide smoother and more stable motion.

Live Z-Adjustment: Real-time adjustment of the nozzle distance from the print bed during the first layer.

Magigoo: A popular 3D printing adhesive that helps with bed adhesion.

Nozzle: The component through which the filament is extruded onto the print bed during printing.

Offset: Adjusting the position of the print head to fine-tune the print's dimensions.

Over-Extrusion: The printing of too much material, resulting in bulging and dimensional inaccuracies.

Overhang: A portion of a 3D print that is printed at an angle without any support underneath.

Perimeter (Shell): The outermost layer of a 3D print that defines its shape and provides structural integrity.

PID Tuning: The process of calibrating the heating element to maintain a stable and accurate temperature during printing.

Printhead Cooling: The cooling fan or system used to cool the print head and nozzle during printing.

Print Bed: The surface upon which the 3D printed object is built.

Print Head: The assembly that includes the nozzle and heater block for extruding filament.

Print Speed: The speed at which the print head moves during printing.

Raft: A base layer of printed material used to improve bed adhesion and reduce warping.

Retract and Prime Tower: A structure printed alongside the main object to minimize stringing by retracting and priming the filament when moving between different parts of the print.

Retraction: The process of pulling filament back into the nozzle to prevent oozing during non-printing moves.

Simplify3D: A commercial slicing software known for its advanced features and capabilities.

Slicer: Software that converts 3D models into G-code instructions for 3D printing.

Slic3r: An open-source slicing software for 3D printing.

Shell Thickness: The number of perimeters (shells) on the outside of the print.

Spool Holder: A device that holds the filament spool and feeds it into the printer.

Spooling: The process of unwinding the filament from the spool during printing.

STL: Standard Triangle Language - A common file format used for 3D models in 3D printing.

Stepper Motor: Motors used in 3D printers to control precise movements of the printer's components.

Support Structures: Additional structures printed to support overhangs and prevent sagging during printing.

Tolerance: The acceptable margin of error in the dimensions of a 3D print compared to the original 3D model.

Travel Speed: The speed at which the print head moves when not extruding filament.

Z-Hop: The vertical movement of the print head during travel moves to avoid collision with printed parts.

Z-Seam: The location where the printer starts a new layer, affecting the appearance of the print.

Z-Wobble: Visible patterns or lines on the print surface due to issues with the Z-axis movement.

Citations

1. Thangs. (2023, August). Retrieved from thangs.com

2. Sketchfab. (2023, August). "Introduction to 3D Modeling." Retrieved from sketchfab.com

3. STLFinder. (2023, September). "Guide to Finding the Best STL Files." Retrieved from www.stlfinder.com

4. TurboSquid. (2023, September). "3D Models for Professional Use." Retrieved from www.turbosquid.com

5. Yeggi. (2023, September). Retrieved from www.yeggi.com

6. Free3D. (2023, September). "Diverse 3D Models for Printing." Retrieved from free3d.com

7. Pinshape. (2023, October). "Learning 3D Printing Through Community." Retrieved from pinshape.com

8. YouMagine. (2023, October). "Collaborative 3D Printing Designs." Retrieved from www.youmagine.com

9. Tinkercad Blog. (2023, November). "Easy 3D Design for Beginners." Retrieved from tinkercad.com

10. Cults. (2023, November). "Design Contests and Community." Retrieved from cults3d.com

11. 3DPrint.com. (2023, November). "Latest Trends in 3D Printing." Retrieved from 3dprint.com

12. 3D Printing Industry. (2023, November). "In-Depth Industry Analysis." Retrieved from 3dprintingindustry.com

13. All3DP. (2023, December). "Comprehensive Guides on 3D Printing." Retrieved from [all3dp.com](all3dp.com)

14. Make Magazine's 3D Printing Section. (2023, December). "Innovative 3D Printing Projects." Retrieved from makezine.com

15. Thingiverse. (2023, December). "Expansive Repository of 3D Designs." Retrieved from www.thingiverse.com

Notes

Here's a place to use as your own and keep handy through your printing journey. A few ideas for this space would be:

- A maintenance log
- Tested material print settings
- Ideas for your next parts to print

Good luck and happy printing!

_____

_____

_____

_____

_____

_____

_____

_____

_____

_____

_____

_____

72

_____

_____

_____

_____

_____

_____

_____

_____

_____

_____

_____

_____

_____

_____

_____

_____

Page intentionally left blank

The Hobbyist's Guide to 3D Printing

fin

www.ingramcontent.com/pod-product-compliance
Lightning Source LLC
Chambersburg PA
CBHW062241290526
45794CB00006B/2362